P9-EKW-955

TRUE or FALSE?

BIOHAZARD

This cat can be extremely dangerous.

TRUE!

Don't be fooled by this cute face. Some kittens carry germs that cause an illness called cat scratch fever. (It's also known as cat scratch disease.) Fever, chills, headaches, and stomachaches are some of the symptoms. A scratch or bite from this feisty little critter could make you sick for months!

Book design Red Herring Design/NYC

Library of Congress Cataloging-in-Publication Data
Brownlee, Christen, 1977–
Cute, furry, and deadly : diseases you can catch from your pet! / by Christen Brownlee.
p. cm.—(24/7 : science behind the scenes)
Includes bibliographical references and index.
ISBN-13: 978-0-531-12072-9 (lib. bdg.)   978-0-531-18737-1 (pbk.)
ISBN-10: 0-531-12072-4 (lib. bdg.)   0-531-18737-3 (pbk.)
1. Zoonoses—Juvenile literature. 2. Communicable diseases in animals—Juvenile literature. 3. Animals as carriers of disease—Juvenile literature. 4. Pets—Diseases—Juvenile literature. I. Title.
RA639.B76 2007
614.4'3—dc22   2006021230

CUTE, FURRY, AND DEADLY

Diseases You Can Catch From Your Pet!

Christen Brownlee

WARNING: The cases in this book are true stories. They all involve adorable animals who made people deathly ill. Read at your own risk.

Franklin Watts
An Imprint of Scholastic Inc.
New York • Toronto • London • Auckland • Sydney
Mexico City • New Delhi • Hong Kong
Danbury, Connecticut

CONTENTS

MEDICAL 411

Claw through this section, and get the 411 on the world's most dangerous pets.

This pet will make this Wisconsin woman sick.

TRUE-LIFE CASE FILES!

Find out how zoonotic researchers solved some major medical mysteries.

15 Case #1:
The Case of the Puzzling Prairie Dog

An unusual pet makes a girl and her mother sick. Can doctors discover what disease the pet has spread?

27 Case #2:
The Case of the Killer Kitten

A litter of kittens is the hit of the pet fair. But did one family take home a killer?

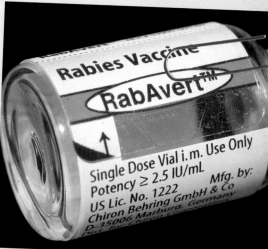

In Maryland, someone needs the rabies vaccine.

37 Case #3:
The Case of the Fearful Festival

This fair was no picnic. Why did so many kids end up at the hospital after visiting the Strawberry Festival?

In Florida, a peaceful fair becomes—for some—a nightmare.

5

Hopefully, the worst thing your pet ever comes home with is bad breath. But some furry friends can carry nasty germs. And some of these germs can kill people—without even making the animals sick.

MEDICAL 411

That's where zoonotic disease researchers come in. They investigate and stop diseases that come from animals—before these diseases stop you!

IN THIS SECTION:

- ▶ how zoonotic disease researchers really talk;
- ▶ germs that pets can carry—and what these germs can do to you;
- ▶ and people who work to prevent and treat diseases caused by animals.

Animal Talk

Zoonotic disease researchers have their own way of speaking. Find out what their vocabulary means.

zoonotic
(zoh-uh-NOT-ik) having to do with a disease that's passed from animals to people. Wild animals, pets, and livestock can all cause *zoonotic* illnesses.

The patient has a rare disease. I suspect it's **zoonotic**.

We need to find out where the carrier is so we can stop the disease from spreading.

carrier
(KA-ree-ur) a host that can pass on disease-causing bacteria or viruses but usually doesn't get sick itself

symptoms
(SIMP-tuhmz) health conditions that indicate illness

I'm not familiar with these symptoms. We're dealing with a very rare disease.

Take some blood, and we'll run tests to figure out what **bacteria** or **virus** is making the patient sick.

Say What?

Here's some lingo an expert in zoonotics might use on the job.

bacteria
(bak-TIHR-ee-uh) tiny, living, single-cell creatures that exist all around and inside you. Many are useful, but some cause diseases.

virus
(VYE-ruhss) a very tiny living thing that can reproduce and grow only when inside living cells. A virus can cause disease.

culture
(KUHL-chur) bacteria or viruses growing in petri dishes or test tubes
*"We'll let the **culture** grow for a day or two before running tests on it."*

host
(hohst) an animal or plant from which another organism gets nutrition
*"This virus always kills its **host** after a week or two."*

vaccine
(vak-SEEN) an injection of a weakened or dead bacteria or virus. A vaccine is given to prevent or treat infectious diseases.
*"Just to be safe, she got the rabies **vaccine** after a stray dog bit her."*

I'm not going home until we **diagnose** this problem.

diagnose
(dye-uhg-NOHSS) to identify a disease or illness by a medical examination and tests

What's Living on Your Pet?

Here's a look at some of the germs that can move off of animals—and onto humans!

Billions of bacteria and viruses are living in and on your body right now! And cats, dogs, and other pets have their own set of germs as well.

Most of these germs are completely harmless. But a few can cause serious illnesses. Check out the disease-causing bacteria or viruses that some animals carry. And find out what they can do to you!

Disease: **Toxoplasmosis**
Most common carrier: **Cat**
Microorganisms: ***Toxoplasma gondii***
Symptoms: **Muscle aches and pains, fatigue, swollen glands**

Disease: **Anthrax**
Most common carrier: **Sheep**
Microorganisms: ***Bacillus anthracis***
Symptoms: **High fever, cough, sore throat, headache, nausea**

Disease: **Rabies***
Most common carrier: **Raccoon**
Microorganisms: *Rhabdovirus*
Symptoms: **Anxiety, muscle spasms and weakness, drooling**

* *Rabies can also be carried by other animals, such as dogs, cats, and ferrets.*

▲

▲

Disease: **Bubonic plague**
Most common carrier: **Rat**
Microorganisms: *Yersinia pestis*
Symptoms: **Chills, fever, diarrhea, headaches, swollen glands**

Disease: **Tularemia**
Most common carrier: **Rabbit**
Microorganisms: *Francisella tularensis*
Symptoms: **Fever, headache, cough, diarrhea**

▽

▲

Disease: **Hantavirus**
Most common carrier: **Mouse (and other rodents)**
Microorganisms: *Hantavirus*
Symptoms: **Fever, trouble breathing, heart failure**

11

The Medical Team

Zoonotic disease researchers work as part of a team. Here's a look at some of the people who help solve medical mysteries.

EPIDEMIOLOGISTS
They study the factors affecting the health of a community. Their job is to control and prevent diseases.

LAB TECHNICIANS
They assist the clinical pathologists. They prepare the lab equipment, perform routine tests, and keep the lab sterile.

INTERNAL MEDICINE DOCTORS
They examine the patients and record their symptoms. If they suspect a zoonotic disease, they may consult with other experts.

CLINICAL PATHOLOGISTS
They examine patients' blood, body fluids, or other samples. They try to determine if patients have zoonotic diseases—and if so, what kind.

DERMATOLOGISTS
They are doctors who treat and study skin problems.

INFECTIOUS DISEASE SPECIALISTS
They specialize in diseases caused by parasites, bacteria, viruses, and fungi.

VETERINARIANS
They diagnose and treat ill animals, including those with diseases that can make people sick. They also run tests to find out if an animal is a carrier of a disease.

TRUE-LIFE CASE FILES!

24 hours a day, 7 days a week, 365 days a year, zoonotic disease specialists are tracking down mysterious diseases carried by animals.

IN THIS SECTION:

- ▶ did a unique pet pass on an unusual disease?
- ▶ is a cute little kitten a killer?
- ▶ how did dozens of kids at a fair get so sick?

These three case studies are true. However, some names, places, and other details have been changed.

How do zoonotic disease specialists get the job done?

What does it take to figure out what is wrong with a sick person or animal? Good zoonotic disease specialists don't just make guesses. They're scientists. They follow a step-by-step process to discover what the problem is.

As you read, keep an eye out for the icons below. They'll clue you in to each step along the way.

THE QUESTION At the beginning of a case, the disease specialists have **one or two main questions** they need to answer.

THE EVIDENCE The next step is to **gather and analyze evidence**. What is the person's medical history? What kind of contact with animals has he or she had recently? Disease specialists gather as much information as they can. Then they study it to figure out what it means.

THE CONCLUSION Along the way, disease specialists come up with theories about what may have happened. They test these theories. Does the evidence back up the theory? **If so, they've reached a conclusion**. If all goes well, they'll be able to treat the patient.

WARNING
You are entering an area where animal wastes may be present on animals and surfaces. Microbes in these wastes can cause diarrhea, cramps, nausea, headaches, or other symptoms. They may pose special health risks for infants, young children, some of the elderly, and people with compromised immune systems.

IMPORTANT NOTICE
You are entering an animal area. For the health and safety of all patrons, especially children, please finish all food and drink before entering.

There is No Smoking in this animal area.

Strollers should not be brought into this animal area.

Pacifiers, small toys, spill-proof cups

Marshfield, Wisconsin
May 11, 2003

The Case of the Puzzling Prairie Dog

An unusual pet makes a girl and her mother sick. Can doctors discover what disease the pet has spread?

Strange Pet, Strange Sickness

A little girl arrives at the local hospital with a very strange set of symptoms.

On Mother's Day 2003, three-year-old Schyan Kautzer and her parents headed to a pet swap in Wausau, Wisconsin. A pet swap is a place to see and buy pets.

The family lives on a farm in nearby Duluth. They have lots of animals. But they always have room for more.

The pet swap had plenty of familiar and unusual new pets. There were gerbils, and there were camels. If something caught the family's attention, they were ready to buy.

Schyan's mother, Tammy Kautzer, had pet prairie dogs when she was younger. The sandy-colored critters were adorable. They cost $95 each. Tammy paid for two. Then, Schyan and her parents loaded the pups into boxes and drove them to their new home.

Schyan loved playing with her cute new pets. But

Steve Kautzer holds three-year-old Schyan, as his wife Tammy holds their prairie dog, Chuckles. The Kautzers' other prairie dog died just after it bit Schyan.

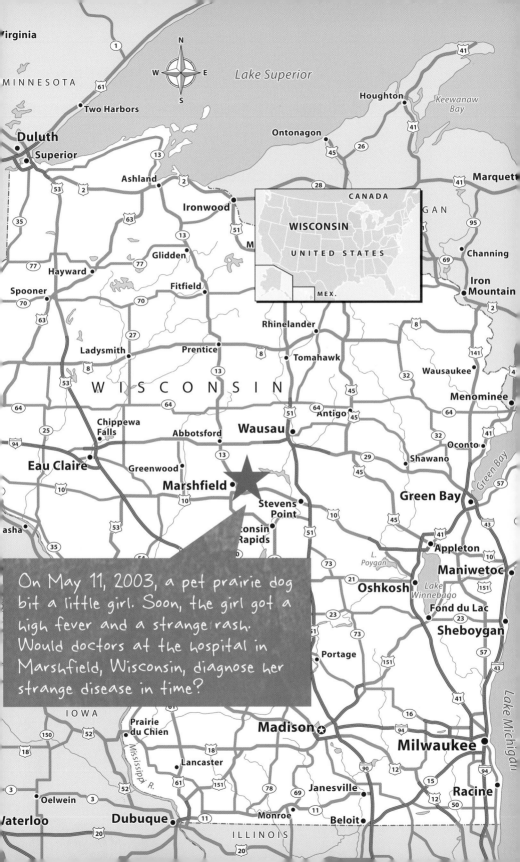

On May 11, 2003, a pet prairie dog bit a little girl. Soon, the girl got a high fever and a strange rash. Would doctors at the hospital in Marshfield, Wisconsin, diagnose her strange disease in time?

soon, one of the prairie dogs started looking sick. It moped around. Its eyes looked crusty. It seemed cranky, just like a sick person.

Schyan picked up the pup to put it back in its cage. It gave her finger a nasty bite. Tammy and Schyan's father, Steve Kautzer, washed the bite and kissed it.

Over the next few days, the bite mark looked like it was healing. But then the bite turned into a dark red bump. More unusual-looking bumps popped up nearby. And Schyan came down with a high fever.

The sick prairie dog died.

Worried that Schyan might be seriously ill, her parents drove her to the hospital in Marshfield, Wisconsin.

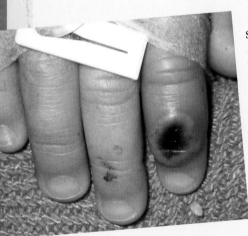

A child's infected finger two weeks after being bitten by a prairie dog. After Schyan Kautzer was bitten by her pet, she developed a large bump on her finger like this one.

Medical Mystery

Doctors search for information about Schyan's sickness. Could the bumps on her body hold clues?

Dr. Scott David came into the hospital room to talk with Schyan's parents. The girl would need a few tests to diagnose her disease. He needed to know what was causing her infection before he could fight it.

He took samples of Schyan's blood. Then he asked her parents lots of questions. Had she had anything strange to eat or drink? Had she been around other kids? Had anything unusual happened in the past few days?

Tammy mentioned that her daughter had been nipped by a sick prairie dog. They'd thrown the dead critter away before driving to the hospital.

The doctor told Tammy to dig the pup's body out of the trash and bring it back to the hospital.

Prairie dogs are rodents. Rodents can carry many different diseases. The animal's body needed to be studied and tested. That might give the doctors clues about what was making Schyan sick.

Schyan stayed in the hospital for the next three days. Doctors ran test after test. Still, they couldn't figure out what was wrong. The

Had Schyan's prairie dog made her sick? After all, prairie dogs are rodents, and rodents carry a lot of diseases.

doctors gave her several different **antibiotics**. But none made her feel better.

Schyan's doctors realized they needed some help. They called Dr. John Melski, a **dermatologist** and medical information expert. He helps solve mysterious skin diseases when other doctors are stumped.

Digging for Clues

The usual drugs don't work on this unusual illness. A new investigation begins.

Dr. Melski's home phone rang on a Sunday morning. He listened as a doctor at the hospital listed Schyan's symptoms. She had a high fever, loss of appetite, and **fatigue**. She also had strange, blister-like bumps that now covered her entire body.

Schyan's illness was getting worse. "They needed my help pretty quickly," Dr. Melski says.

What was making Schyan so sick? What was causing her strange symptoms? Dr. Melski needed to find out fast.

Dr. Melski decided to look up some information before he left his house. He logged on to the Internet. He read about diseases that prairie dogs can carry. But none of those diseases cause symptoms like the ones Schyan had.

He hopped in his car for the mile-long (1.6-km) drive to the hospital. As soon as he arrived, Dr. Melski read notes that the other doctors had taken.

Dr. Melski noticed that Schyan had received antibiotics for several diseases that cause skin rashes and fevers. But the drugs hadn't eased her symptoms.

"I realized that this was a strange case," Dr. Melski says. "Most of the diseases I uncovered should have been treated by the antibiotics, but Schyan wasn't getting better."

When Dr. Melski examined the bumps on the girl's hands, he got his first clue about the disease. Some of the fluid-filled bumps had dimples in the center.

"My instincts told me that what I was seeing was viral," he says. Other infections caused by viruses, such as chicken pox or cold

21

Dr. Melski analyzed a sample of fluid from Schyan's wound under a microscope. He found nothing. Then he sent a sample of her diseased skin to the hospital lab.

THE EVIDENCE

sores, cause bumps with a similar appearance.

Dr. Melski knew that the bumps themselves held more clues. He put on gloves and cut open one of the blisters. He drained the fluid onto a glass slide. Then, he added a stain that sticks to viruses to make them easier to see under a microscope. But Dr. Melski saw nothing.

Hunting down the virus would take more drastic measures. He needed to do a **biopsy**.

First, he numbed the skin on Schyan's hand. Then, Dr. Melski used a blade to cut out a bump the size of a pencil eraser. He sent the skin sample off to the hospital's lab.

Lab technicians cut this sample into thin slices and placed them on glass slides. Then a **pathologist** named Dr. Kurt Reed took a look at the slides.

Pathologists like Dr. Reed are trained to examine samples for signs of disease. Dr. Reed told Dr. Melski he'd been right. Schyan had a virus.

But which virus? They still didn't know.

Picture Power

Now, Schyan's mother falls ill. And the doctors still don't know what the mystery disease is.

Back at the hospital, Schyan was feeling much better. Her fever was gone. Her appetite was back. But now her mother was sick. Blister-like bumps were forming around a cut on Tammy's hand. She also had a fever.

Dr. Melski took a skin sample from Tammy. They also snapped some pictures of the viruses so Dr. Melski could show other doctors.

Dr. Melski shared the images with Dr. Reed. They were almost certain that the pictures showed a pox virus. That's a virus that causes raised, infected bumps.

Still, there are many different types of pox viruses. How could the doctors figure out which virus this was?

Over the next few days, other cases of the strange pox started popping up in Milwaukee. These patients had also bought prairie dogs at the pet swap.

It was time to call the Centers for Disease Control and Prevention (**CDC**). That's a government organization in charge of protecting public health. They try to track and control **contagious** diseases.

Schyan's medical team posted photos of the mystery virus on the Internet to get help identifying it. The virus turned out to be monkeypox, which is shown here.

Dr. Melski and other doctors working with the pox patients met with officials from the CDC. The CDC officials told the doctors to post pictures of the mystery pox virus on the Internet. That way, doctors around the world could share their knowledge.

THE CONCLUSION Within hours of posting the photos online, Dr. Melski finally had a diagnosis. His patient had a virus called **monkeypox**.

Out of Africa

How did a deadly virus travel half a world away? Doctors solve the mystery.

Monkeypox! The diagnosis was a surprise. This illness is most common in central and western Africa. It's called monkeypox because it was first found in monkeys scientists were studying in a lab. But African squirrels are probably the most common carriers of the disease.

How did the disease travel to the U.S.? Scientists from the CDC wanted to find out.

After investigating, they learned that the prairie dog pups had been kept next to a rat at some point. The rat was from Gambia, a country in Africa.

These Gambian rats can be carriers for monkeypox. They can look perfectly healthy but make other animals sick.

In total, 54 people ended up with monkeypox infections from the pet prairie dogs, including Schyan's father, Steve. All the patients survived. But Dr. Melski explains that giving the right care to patients was important for their recovery. People can die from monkeypox if they don't get medical care (see page 26).

After figuring out what disease had struck Schyan, doctors were able to treat the other patients. Schyan and her family have since made a full recovery. And Dr. Melski and other doctors know what to look for if there's ever another monkeypox outbreak.

Monkeypox isn't a typical disease carried by prairie dogs, says Dr. Melski. "That's why I missed it in my initial research." He adds, "Next time, we'll be more prepared." **24/7**

Schyan's prairie dog had been housed next to a Gambian rat like this one. These rats can be carriers for monkeypox.

Dr. John Melski talks about the monkeypox case—and what it's like to solve a medical mystery.

24/7: Why did you become involved in finding Schyan's diagnosis?

DR. JOHN MELSKI: I am a consultative dermatologist. I solve puzzles relating to the skin.

24/7: How important was working with other disease experts in solving this mystery?

DR. MELSKI: The CDC is an incredible resource. Officials there helped us figure out that we were dealing with monkeypox.

24/7: How do doctors treat monkeypox?

DR. MELSKI: Schyan and most other patients got what we call supportive care. We made sure they had enough nutrients and fluids. But their **immune systems** took care of the recovery.

24/7: What if the patients hadn't had supportive care?

DR. MELSKI: Monkeypox can be deadly. In African countries, where people get this disease more frequently, patients without medical care sometimes die.

In this case, doctors had to hunt down a strange disease. But how do researchers handle a well-known—but deadly—disease?

The Case of the Killer Kitten

A litter of kittens is the hit of the pet fair. But did one family take home a killer?

27

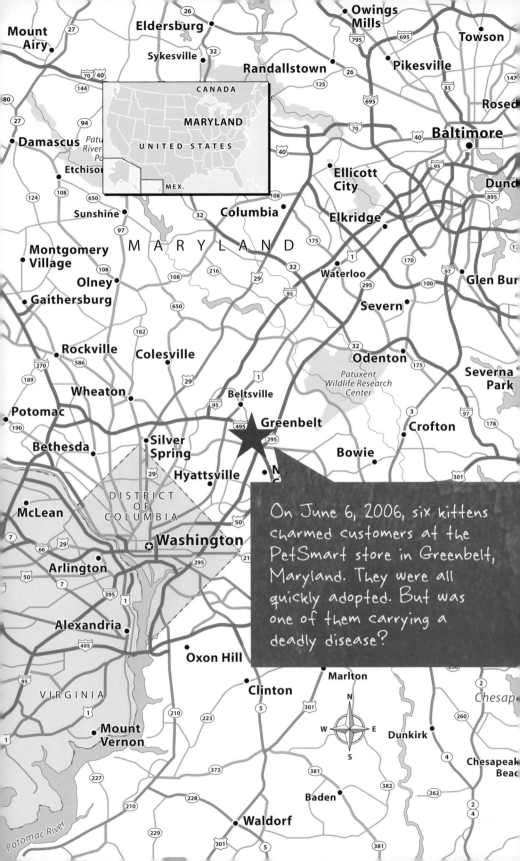

On June 6, 2006, six kittens charmed customers at the PetSmart store in Greenbelt, Maryland. They were all quickly adopted. But was one of them carrying a deadly disease?

Adorable and Adopted

Rescued kittens all find new families. But one is acting very strangely.

On a bright spring day in Aquasco, Maryland, the phone rang at Last Chance Animal Rescue. It was a typical call. The caller wanted to find homes for stray animals.

This time, it was a litter of kittens. They were living under the porch of the caller's house. Could the group come by and pick them up?

Last Chance's director, Cindy Sharpley, rounded up some volunteers and headed over to the house. They loaded all six of the ten-week-old kittens into a truck. Then, they headed back to the group's headquarters.

The kittens were adorable, thought Sharpley. They'd get snatched up quickly at the group's next adoption fair. The fair would be held at a PetSmart store in Greenbelt, Maryland.

A litter of kittens was brought to a PetSmart store for adoption. All the kittens got new homes within a matter of days.

Kittens—as well as many other mammals—can carry dangerous diseases.

Before the fair, a **veterinarian** examined all the kittens. She said they were all healthy. Then, just as Sharpley expected, the kittens stole the show at the fair. Lots of visitors stopped to pet them, pick them up, and cuddle them.

Over the next week, all the kittens found new families. The new owners filled out paperwork to take their new pets home.

But within days, Sharpley got a surprising call from one of the new owners. He wanted to know if he could return his kitten. It was acting unlike any kitten he'd ever seen.

Killer Bite

Even a tiny nip can pass on the kitten's deadly infection.

The owner brought the kitten back to Last Chance. It nipped Sharpley's hand when she reached into the cage. The owner said that his new pet had been very aggressive. It had bitten him as well. The kitten was making strange growling sounds. It also was staggering around the cage.

Sharpley knew that the whole litter had received a clean bill of health just a couple weeks before. Still, the kitten was obviously very sick. What disease could possibly have come on this quickly?

She called in a vet to take a look.

The vet examined the kitten. She observed how it was acting. Then she told Sharpley that she suspected the kitten had **rabies**.

The kitten was probably already carrying the virus when it was first examined. Rabies symptoms can take weeks—or even months—to appear. But once symptoms begin, certain death is near.

Rabies is caused by a virus, explains rabies expert Kim Mitchell of the Maryland Department of Health and Mental Hygiene in Baltimore. Only mammals can get rabies. Mammals are animals that have hair and make milk. So that means that dogs, cats, and people can get infected, but snakes, turtles, and fish can't.

The rabies virus is only present in infected animals' saliva. (Saliva is spit.) It's almost always **transmitted** through a break in the

Before the kittens were adopted, they were examined by a veterinarian. The Maryland kitten was probably already carrying rabies when it was examined. But it didn't show any symptoms.

31

skin, such as a bite. "People can't get rabies by petting or hugging animals or even by getting an animal's blood on their skin if there is no wound," says Mitchell.

But rabies is a very difficult disease to control. The disease itself makes animals very aggressive. They'll bite, scratch, and attack—even people they know and love. And as they do that, they spread the deadly virus.

How does this killer work? The rabies virus attacks the nerves of the body first. Later, it travels to the brain. It causes fever, muscle weakness, drooling, and behavior changes. Animals and people with rabies almost always die of swelling of the brain.

Luckily, there are vaccines that can protect humans. Pets should be vaccinated *before* any **exposure** might occur. And people can be given special shots after they are bitten by a strange or wild animal. The shots must be given *before* any rabies symptoms show up in the person. This treatment isn't fun. But it saves many lives every year.

Would Sharpley and the kitten owner need the vaccine? The investigation continued.

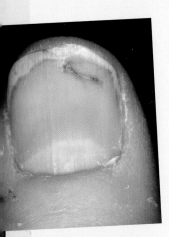

This is a cat bite on a human finger. Rabies is spread when animals bite or scratch people.

The Diagnosis

Researchers can't tell if an animal definitely has rabies—unless they look at its brain.

The kitten definitely had symptoms of rabies. But Sharpley couldn't tell for sure if the kitten had rabies unless it was tested for rabies.

Unfortunately, to test an animal for rabies, it has to be **euthanized**. That means it's killed in a painless way.

Sharpley sent the euthanized kitten to the state laboratory. There, lab workers sliced the brain tissue into thin sections and put them on glass slides. The researchers then added a dye that sticks to the rabies virus.

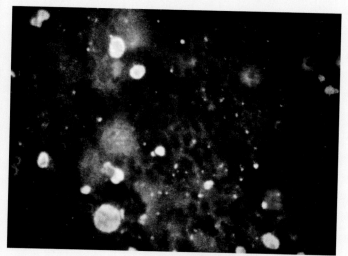

The green, glowing, bullet-shaped objects are the rabies virus, as seen through a microscope.

Five different shots during a one-month period. That's what's given to patients who may have rabies.

Next, they placed the slides into a specially designed machine that makes the dye glow green.

THE CONCLUSION The lab workers had their proof. Glowing circles lit up the samples of the kitten's brain. That was a certain sign that it had rabies.

Sharpley and the kitten's owner quickly started the series of lifesaving vaccine shots.

But there was another question to answer: Was this the only kitten in the litter with the disease? State officials couldn't take any chances. All of the rabid kitten's siblings had to be rounded up, euthanized, and tested for rabies. They all turned out to be disease-free.

RabAvert is the vaccine for people who have been bitten by an animal suspected of having rabies. It is made from a weakened version of the virus.

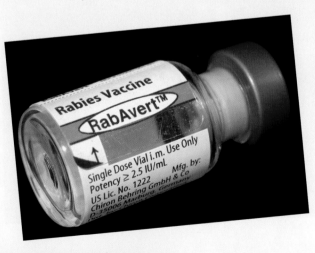

Calling All Kitten Lovers

Did anyone else have contact with the rabid kitten?

The danger wasn't over, however. The sick kitten had been on display for a week. Anyone who had stopped by the adoption fair could have been exposed to rabies. What if someone else had been infected?

Workers at the health department interviewed PetSmart employees. Had anyone been bitten by the kittens?

The health department contacted newspapers, radio, and TV stations. They wanted everyone who stopped by the fair to know they might have been exposed to rabies. They also told local doctors and hospitals about what had happened.

"It was a massive amount of people we were trying to reach," says Mitchell.

Their efforts paid off. No one developed rabies because of the pet fair.

Finding rabies in an animal up for adoption is extremely rare. But Mitchell notes that the experience has health department officials on alert for future cases.

"Vigilance is key to prevention," she says. 24/7

Kim Mitchell explains the painful facts about rabies.

24/7: Are some mammals more likely to carry rabies than others?

KIM MITCHELL: Yes, but it depends on location. Here in Maryland, raccoons lead the pack. For states in the Midwest, it's bats and skunks.

24/7: Durning which seasons are people more likely to get exposed to rabies?

MITCHELL: People can get exposed pretty much year-round, but we tend to see more bites by animals that may be rabid during the summer months.

24/7: Our pets can get vaccines that prevent them from getting rabies if they get bitten. What about people?

MITCHELL: Once a person is bitten, he or she usually needs a series of five shots to stop the disease. That's called the post-exposure series. People who work with animals, like veterinarians, can get a pre-exposure vaccination. Then, if they get bitten, they only need two shots.

24/7: Is there any good way to protect yourself from being exposed to rabies?

MITCHELL: It's fine to play with your own pets, but leave other pets and animals alone.

House pets aren't the only animals that spread diseases. Can investigators identify what made kids sick at a local fair?

Plant City, Florida
March 13, 2005

The Case of the Fearful Festival

This fair was no picnic. Why did so many kids end up at the hospital after visiting the Strawberry Festival?

Festival Fallout

Doctors identify a killer bacteria. But where did it come from?

It was a typical day at the Florida Strawberry Festival for six-year-old Shannon Smowton and her family. Like hundreds of other kids at the fair, Shannon went on rides, watched parades, and ate loads of sugary food. She also ate lots of strawberries.

A few days after the festival, Shannon got extremely sick. She had some alarming symptoms. She had bloody **diarrhea**, a high fever, and the beginnings of kidney failure.

Her parents rushed her to the hospital.

Shannon wasn't the only kid in central Florida with these symptoms. Dozens of kids

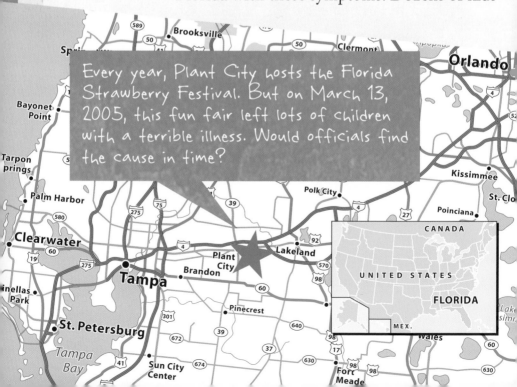

Every year, Plant City hosts the Florida Strawberry Festival. But on March 13, 2005, this fun fair left lots of children with a terrible illness. Would officials find the cause in time?

were going to local hospitals with these same problems.

Tests showed that the kids all had a very dangerous **strain** of a bacteria called **E. coli.** It's so dangerous that doctors are ordered to report cases to the Florida state health department.

Health department officials quickly concluded that they had an epidemic on their hands. That's when a disease affects many people at the same time.

They also had a mystery. This dangerous strain of E. coli is pretty rare. Why did all these children from the same area get this disease at the same time?

Dr. Roberta Hammond is an epidemiologist at the Florida State Health Department. It's her job to track down diseases in the community. Then she tries to keep these diseases from spreading.

What strange disease was lurking at the Florida Strawberry Festival?

Unfair Fair

Most kids who attended the fair didn't get sick. So what do the sick ones have in common?

Dr. Hammond and other workers from the health department began their investigation.

They knew that all the kids had attended the Strawberry Festival and other nearby fairs. But what else did these children have in common?

"You want to know how they got it—what made them sick," says Hammond.

First, Dr. Hammond and the other workers went to the hospitals to visit the sick children and their parents. These researchers wanted to talk with the parents about their kids' experiences at the Strawberry Festival.

The researchers started their interviews by asking what the children had eaten at the fair. That's because E. coli often comes from contaminated food—especially contaminated meat.

But after questioning the parents, the scientists were still stumped. Many of the children had eaten meat, but some hadn't. In fact, their diets that day had little in common.

Dr. Hammond recalls, "Nobody ate the same food. Not only that, but the foods they ate were mostly ones that we don't normally think of as making people sick. We had to look in other directions."

Eventually, researchers hit on one important question: Had the children been in contact with animals at the fair?

THE EVIDENCE

All the families had the same answer: Yes! The sick children had all gone into petting zoos at the fairs. They'd petted baby cows, goats, and sheep. And many kids had also fed the animals.

THE QUESTION

Could these animals be the source of the kids' sickness?

Lots of kids at the Strawberry Festival spent a lot of time with the animals at the petting zoo. Did they pick up a disease while they were there?

Pointing Fingers

Researchers can't blame the petting zoos without proof. Could a high-tech test hold the answer?

Hammond's team suspected that the petting zoos might be hosting deadly E. coli. But the researchers didn't want to blame the zoos without being 100 percent sure.

To check their suspicions, the researchers had to test the animals and their pens. They ran swabs over the animals and around the pens. They placed these swabs in **sterile** bags to make sure that other bacteria didn't get into the samples.

Back at the lab, the researchers wiped the swabs on **petri dishes**. That way, any bacteria on the swabs would grow.

The dishes grew thick bunches of bacteria.

Hammond's team then ran tests to see if any of these bacteria matched the deadly E. coli strain that infected the kids.

This is an enlarged image of E. coli bacteria.

Sure enough, the bacteria from the animals matched the bacteria that caused the kids' infections. The epidemiologists had found the source of the dangerous E. coli.

Washing and Waiting

Researchers come up with a squeaky-clean plan to prevent future outbreaks.

Eventually, all the children recovered—even though many of them had been extremely ill.

Still, Dr. Hammond and others at the health department were concerned. How could they make sure that the zoos wouldn't make kids sick again?

Officials at the health department made a difficult decision. They had the animals that carried the dangerous bacteria euthanized. That way, these animals wouldn't make any other people sick.

These days, visitors to petting zoos are encouraged to wash their hands after visiting the animals.

Florida officials also published suggestions for petting zoos. They said that all petting zoos should supply hand-washing stations. That way, anyone who touches the animals could clean up before leaving.

"One of the best things you can do to keep people healthy is to educate them about hand washing," says Dr. Hammond.

Your mom was right. The best way to stay healthy *is* to wash your hands. **24/7**

Dr. Roberta Hammond talks about why kids got sick with E. coli—and how you can stay healthy.

24/7: Where does this dangerous strain of E. coli usually live?

DR. ROBERTA HAMMOND: E. coli O157:H7 lives in some animals' intestines and comes out in their manure. Usually these bacteria are in undercooked meat. But recently, they have been found in sprouts, spinach, and other foods.

24/7: Why do you think the kids at the fair got infected?

DR. HAMMOND: The bacteria were on the animals. The kids could have petted the animals and then put their hands in their mouths. That would have put the dangerous bacteria right inside their bodies.

24/7: Why not just shut down petting zoos to prevent people from getting infections like E. coli?

DR. HAMMOND: Some health officials suggested doing that. But many people want to keep the petting zoos. They think these zoos teach children about animals.

24/7: Is there a right way to wash your hands?

DR. HAMMOND: Yes! People should wash all surfaces of their hands. That includes the back of the hand and between the fingers. And make sure you wash for at least 20 seconds.

MEDICAL DOWNLOAD

Go wild with more info on the nasty, natural world of animal-borne diseases.

IN THIS SECTION:

- ▶ how a teen survived rabies;
- ▶ zoonotic diseases make headlines;
- ▶ tools that lead to major medical breakthroughs;
- ▶ what it takes to be a disease detective.

1347–1351 Rats! Rodents Carry Plague
  In October 1347, a ship docked in Messina, Italy. It was carrying rats that would spread a disease that became known as the Black Death. The disease killed millions of people throughout Europe.

Key Dates in the History of Zoonotic Disease

Here are some animals that have made people sick—and some discoveries that have kept them healthy.

1885 The First Rabies Survivor
  In France, nine-year-old Joseph Meister was bitten by a dog with rabies. Back then, that situation was usually hopeless. But a scientist named Louis Pasteur (*left*) gave the boy the first vaccine against rabies. It was a weakened form of the virus. That vaccine saved his life.

1964 Germ Fighters Unite

A group of scientists and doctors formed the **Infectious** Disease Society of America. Today, the society's thousands of members work together to help diagnose, prevent, and treat all sorts of contagious diseases, including **zoonoses**.

1986 Mysterious Mad Cows

English farmers noticed that some of their cows were acting, well, crazy. The cows were nervous and staggering. Scientists discovered they had a new illness. The disease got the nickname "mad cow disease." Soon, researchers found that people who ate some body parts of these cattle also got sick from a similar disease.

1993 Southwestern Outbreak

Almost 100 people in New Mexico, Arizona, Colorado, and Utah went to hospitals with similar symptoms. They were coughing, wheezing, and couldn't breathe. About half of them died. Scientists soon found that these people were infected with hantavirus. Researchers traced the disease to mice in the area.

2004 Rabies Breakthrough

In Wisconsin, 15-year-old Jeanna Giese (*right*) was bitten by a bat. She developed symptoms of rabies—which is usually a death sentence. For the rabies vaccine to work, it has to be given before these symptoms appear. But Jeanna's doctors developed a special treatment. And she became one of the only people to survive rabies.

In the News

Read all about it! Zoonotic disease is front-page news.

Baby Chicks Make Kids Sick

SANTA FE, NEW MEXICO—April 30, 2005

Baby chickens sold as pets have made dozens of people sick. The chickens were born in a hatchery in New Mexico. They were shipped to many different places.

In all, 26 people in 15 states came down with **salmonella** infections. More than half of those people were under five years old.

Salmonella causes diarrhea, fever, and stomach cramps. Symptoms show up 12–72 hours after exposure. Most people recover without medication. But small children can get very sick from the disease.

People catch the virus from the baby chick's feces. The most important way to prevent it is to wash your hands well after handling baby chicks.

Officials also say that you should not buy pet chickens if anyone in your household is under five years old.

These fuzzy little chicks can pack a punch. In New Mexico, chicks carried the salmonella virus, and more than two dozen people got sick.

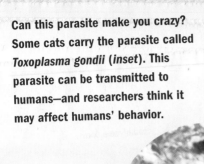

Can this parasite make you crazy? Some cats carry the parasite called *Toxoplasma gondii* (*inset*). This parasite can be transmitted to humans—and researchers think it may affect humans' behavior.

A Personality Parasite?

ANNAPOLIS, MARYLAND—January 28, 2007

A single-celled animal can get inside your brain and makes you take crazy risks. That may sound like science fiction, but it's not. It's part of a new theory about a **parasite** that often lives in cats.

A parasite is an animal that must live on or in another animal to survive. The *Toxoplasma gondii* parasite lives in cats' guts.

This parasite can be passed to people and other animals through cat feces.

Now researchers think that this parasite might affect people's behavior. They've discovered that rats and mice infected with it act crazy. The rodents take risks that make them more likely to be caught by cats.

"Can [the parasite] also affect human behavior? The answer is maybe," says scientist Edward McSweegan. Scientists are just at the beginning stages of this research.

Zoonotic Toolbox

Zoonotic disease researchers work with some simple tools— and some high-tech ones—to ferret out illnesses.

EQUIPMENT

petri dishes and test tubes Doctors and scientists take samples from a patient's or animal's body or body fluids. Then they try to grow bacteria or viruses from these samples inside petri dishes and test tubes. What's the point? Getting a closer peek at these microbes could offer clues into what caused a patient's illness.

tissue sectioner Like a cheese slicer at a deli, this machine cuts samples taken from patients' or animals' bodies into thin slices. Scientists can examine these tissue slices for the presence of zoonotic bacteria or viruses.

microscope Scientists and lab techs use this tool to get up close and personal with zoonotic bacteria and viruses.

stains Dyes cling to bacteria, viruses, and tissue samples to help scientists see them better under the microscope.

RESEARCH TOOLS

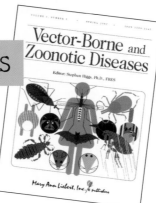

Vector-Borne and Zoonotic Diseases

Editor: Stephen Higgs, Ph.D., FRES

Mary Ann Liebert, Inc./*publishers*

journals Scientists review research recently published by other scientists to get the latest information on zoonotic diseases.

Internet When zoonotic disease researchers are faced with a patient's unrecognizable symptoms, they often search for information on the Internet. Sharing information helps scientists solve their puzzles.

THE OUTFIT

gloves People who study zoonotic diseases often wear gloves to protect themselves from patients' and animals' germs. Gloves also keep lab technicians' hands free from stains and other lab chemicals.

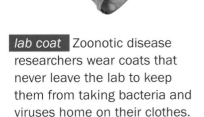

lab coat Zoonotic disease researchers wear coats that never leave the lab to keep them from taking bacteria and viruses home on their clothes.

goggles These protect scientists' eyes from germs, stains, and harmful solutions used in the lab.

51

HELP WANTED:
Dermatologist/Informatician

Ready to put zoonotic disease research under the microscope? Here's some information about this field.

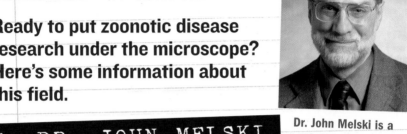

Q&A: DR. JOHN MELSKI

Dr. John Melski is a dermatologist and Medical Director of Clinical Informatics at the Marshfield Clinic in Marshfield, Wisconsin.

24/7: How did you get involved in looking at unusual skin diseases?

DR. JOHN MELSKI: I started out wanting to build computer systems that would help take care of patients. But then I read a really interesting article. I decided that I wanted to learn how to be a doctor from the guy who wrote the article. He was a dermatologist.

24/7: So what is your job, exactly?

DR. MELSKI: I wear two hats. I'm a dermatologist. I help diagnose skin diseases. I'm also an **informatician**. I use computers to analyze medical information and figure out the best care for patients.

24/7: What kind of education did you need to get your job?

DR. MELSKI: I did four years of medical school, then two years of an internship. After that, I did a two-year fellowship in computer-based medicine. Then I did a three-year training program in dermatology.

24/7: What's the hardest part of your job?

DR. MELSKI: There are more things to be done than can be accomplished by one person. There's an endless demand for all the types of jobs that I do. There's a nationwide shortage of dermatologists and informaticians.

24/7: What do you like most about what you do?

DR. MELSKI: I like both sides of my job very much. Taking care of patients is very satisfying on a personal and emotional level. And as an informatician, I'm forever discovering and building tools that will help in the practice of medicine.

24/7: What was your most memorable case?

THE STATS

DAY JOB: Many types of medical professionals and researchers work on zoonotic diseases, ranging from lab technicians to scientists to doctors like dermatologists. These people may work for hospitals, the government, or in private practice.

MONEY: These are the annual salary ranges for some of the jobs in zoonotic disease research.
▶ Dermatologists: $130,000–$420,000
▶ Lab technicians: $20,000–$40,000
▶ Disease researchers (epidemiologists): $40,000–$70,000
▶ Veterinarians: $37,500–$75,000

EDUCATION:
▶ Dermatologists: 4 years of college, 4 years of medical school, 3–5 years of residency.
▶ Lab technicians: High school diploma, plus some college.
▶ Disease researchers: 4 years of college, plus 1–2 years of graduate school.

THE NUMBERS: In the United States last year, there were:
▶ 302,000 lab technicians
▶ 77,000 scientists
▶ 567,000 doctors

DR. MELSKI: That would have to be the monkeypox case. I was quoted in newspapers and all over the place. It was quite a ride. I've worked on lots of cases that were really interesting, but the drama in other cases isn't nearly as publicly recognized.

Take this totally unscientific quiz to find out if zoonotic disease research might be a good career for you.

1 **How are you at figuring out puzzles?**

a) I'm a whiz at figuring out problems. The tougher the better.

b) I will eventually solve them.

c) I'd rather let someone else figure it out.

2 **Are you interested in learning about different kinds of animals?**

a) Yes, I'm fascinated by all kinds of animals.

b) Yes, I love my pets, and I'd like to know more about them.

c) No, animals aren't really my thing.

3 **How do you operate under pressure?**

a) I keep my cool pretty easily.

b) I get nervous sometimes, but I can work through it.

c) The slightest pressure makes me freak out.

4 **Do unusual skin diseases gross you out?**

a) Not in the slightest. So-called gross things are so cool.

b) Gore is not my favorite thing, but I can handle it.

c) Ew, just thinking about that makes me feel sick.

5 **Do you enjoy caring for other people?**

a) Yes, caring for others is one of my core values.

b) I like being responsible for other people sometimes.

c) I'm used to looking out for number one.

YOUR SCORE

Give yourself 3 points for every "**a**" you chose. Give yourself 2 points for every "**b**" you chose. Give yourself 1 point for every "**c**" you chose.

If you got **13–15 points**, you'd probably be a good zoonotic disease researcher. If you got **10–12 points**, you might be a good zoonotic disease researcher. If you got **5–9 points**, you might want to look at another career!

HOW TO GET STARTED...NOW!

It's never too early to start working toward your goals.

GET AN EDUCATION

▶ Starting now, take as many biology, chemistry, physics, health, math, and computer classes as you can. Train yourself to ask questions, gather new information, and make conclusions the way infectious disease specialists do.

▶ Work on your communication skills. Join the drama club or debate team. It's good practice in thinking before you speak and listening to others.

▶ Start thinking about college. Look for ones that have good science and computer programs. Call or write those colleges to get information.

▶ Read the newspapers. Keep up with what's happening in your community.

▶ Read anything you can about infectious diseases. Learn about historical and recent cases. See the books and Web sites in the Resources section on pages 56–58.

▶ Graduate from high school!

NETWORK!

▶ Ask your own doctor for advice about becoming a dermatologist or infectious disease doctor.

▶ Ask a local veterinarian for information about animals and what diseases they try to watch out for.

GET AN INTERNSHIP

Call your local hospital, veterinarian, or doctor's offices. There might be internships available. It doesn't hurt to ask!

LEARN ABOUT OTHER JOBS IN THE FIELD!

parasitologist: studies parasites
biologist: studies living organisms
microbiologist: studies microscopic cells in human illnesses
zoologist: studies animal life
pathologist: studies disease, especially its effects on body tissue
ecologist: studies how organisms relate to the environment infectious disease
pharmacist: dispense medicines

Resources

Looking for more information? Here are some resources you don't want to miss!

PROFESSIONAL ORGANIZATIONS

American Society for Microbiology (ASM)

www.asm.org
1752 N St. NW
Washington, DC 20036-2904
PHONE: 202-737-3600

The ASM supports scientists who study bacteria, viruses, and other microbes, some of which cause zoonotic diseases. Anyone with at least a bachelor's degree in microbiology or a related science can join.

American Veterinary Medical Association (AVMA)

www.avma.org
1931 North Meacham Road
Suite 100
Schaumburg, IL 60173
PHONE: 847-925-8070
FAX: 847-925-1329
E-MAIL: avmainfo@avma.org

The AVMA represents 74,000 veterinarians. This group offers recent updates to its members on news and studies concerning zoonotic diseases.

Centers for Disease Control and Prevention (CDC)

www.cdc.gov
1600 Clifton Road
Atlanta, GA 30333
PHONE: 800-311-3435

The CDC was founded in 1946, primarily to fight malaria. It is part of the Department of Health and Human Services. Today, the group is a leader in efforts to prevent and control disease, injuries, workplace hazards, and environmental and health threats.

Infectious Diseases Society of America (IDSA)

www.idsociety.org
66 Canal Center Plaza, Suite 600
Alexandria, VA 22324
PHONE: 703-299-0200?
FAX: 703-299-0204
E-MAIL: info@idsociety.org

The IDSA represents physicians, scientists, and other health-care professionals who specialize in infectious diseases. The society's purpose is to improve the health of individuals, communities, and society by promoting excellence in patient care, education, research, public health, and prevention relating to infectious diseases.

International Society for Infectious Diseases (ISID)

www.isid.org
1330 Beacon Street, Suite 228
Brookline, MA 02446
PHONE: 617-277-0551
FAX: 617-278-9113
E-MAIL: info@isid.org

The ISID is committed to improving the care of patients with infectious diseases, the training of clinicians and researchers in infectious diseases and microbiology, and the control of infectious diseases around the world.

National Institute of Allergy and Infectious Disease (NIAID)

www3.niaid.nih.gov/
6610 Rockledge Drive, MSC 612
Bethesda, MD 20892
PHONE: 301-496-5717

For more than 50 years, NIAID has conducted research that helps treat, prevent, and better understand infectious and other diseases. It is part of the National Institutes of Health.

Society of Infectious Disease Pharmacists (SIDP)

www.sidp.org/
823 Congress Avenue, Suite 230
Austin, TX 78701
PHONE: 512-479-0425
FAX: 512-495-9031
E-MAIL: sidp@eami.com

The SIDP provides education, advocacy, and leadership in all aspects of the treatment of infectious diseases. The society is comprised of pharmacists and other health-care professionals involved in patient care, research, teaching, drug development, and governmental regulation.

WEB SITES

Mayo Clinic
www.mayoclinic.com

This is a medical site aimed at helping people manage their health.

National Center for Infectious Diseases
www.cdc.gov/ncidod/id_links.htm

Planning to travel to a foreign country? Check out this site to learn how to avoid getting an infectious disease.

National Library of Medicine
www.medlineplus.gov

This is the largest medical library in the world and has information about studies being done on infectious diseases.

Science Magazine Online
www.sciencemag.org

Check out this site's archives for cool parasite articles and photos.

Web MD
www.webmd.com

This is a comprehensive, easy-to-use site with all sorts of helpful medical information.

World Health Organization
www.who.int/en/

This international organization offers infectious disease information in English, French, and Spanish.

BOOKS

Fleisher, Paul. *Parasites: Latching on to a Free Lunch.* New York: Twenty-First Century Books, 2006.

Hayhurst, Chris. *E. Coli* (Epidemics). New York: Rosen Publishing, 2003.

Kiennzle, Thomas E. *Rabies* (Deadly Diseases and Epidemics). Broomall, Pa.: Chelsea House, 2006.

MacTire, Sean P. *Lyme Disease and Other Pest-Borne Illnesses.* Danbury, Conn.: Franklin Watts, 1992.

Smith, Linda Wasmer. *Louis Pasteur: Disease Fighter* (Great Minds of Science). Berkeley Heights, N.J.: Enslow Publishing, 1997.

Yannielli, Yen. *Lyme Disease* (Deadly Diseases and Epidemics). Broomall, Pa.: Chelsea House, 2004.

A

anthrax (AN-thraks) *noun* a disease caused by *Bacillus anthracis*. It is marked by high fever, cough, sore throat, headache, and nausea.

antibiotics (an-tih-bye-OT-iks) *noun* medicines that kill bacteria

B

bacteria (bak-TIHR-ee-uh) *noun* microscopic one-celled organisms. Some bacteria are essential for our survival, while others may cause disease.

biopsy (BYE-op-see) *noun* the process of removing a piece of tissue to examine it

bubonic plague (byoo-BON-ik playg) *noun* a disease caused by *Yersinia pestis*. It is marked by chills, fever, diarrhea, headaches, and swollen glands.

C

carrier (KA-ree-ur) *noun* a host that can pass on disease-causing bacteria or viruses without getting sick

CDC (see-dee-SEE) *noun* a government agency in charge of protecting public health. It is short for the *Centers for Disease Control and Prevention.*

contagious (kon-TAY-juss) *adjective* describing something that is easily transmitted between people or animals

culture (KUHL-chur) *noun* bacteria or viruses growing on petri dishes or in test tubes

D

dermatologist (der-mah-TOL-uh-jist) *noun* a doctor who specializes in skin diseases

diagnose (dye-uhg-NOHSS) *verb* to identify a disease or illness by a medical examination and tests

diarrhea (dye-uh-REE-uh) *noun* a condition that involves frequent and watery bowel movements

Dictionary

E

E. coli (ee KOHL-eye) *noun* a species of bacteria usually found in the intestines of humans and other animals. It is short for *Escherichia coli*. While internal E. coli is harmless and helpful in digestion, eating or drinking E. coli that comes from outside (like polluted water or meat that's not processed safely) can cause severe food poisoning or death.

euthanized (yoo-thah-NIZED) *verb* killed in a painless way

exposure (ek-SPOH-zur) *noun* the state of being in contact with illness while being unprotected

F

fatigue (fuh-TEEG) *noun* extreme tiredness

H

hantavirus (HON-tuh-VYE-russ) *noun* a bacterial disease spread by rodent feces and urine. It can cause breathing difficulties and fever.

host (hohst) *noun* the person or animal in which disease-causing bacteria or viruses live. These microbes may or may not make hosts sick.

I

immune systems (ih-MYOON SIS-tuhmz) *noun* the systems that protect a body against disease and infection

infectious (in-FEK-shuhss) *adjective* capable of spreading disease

infectious disease specialists (in-FEK-shuhss duh-ZEEZ SPESH-uh-lists) *noun* scientists who are experts in infectious and deadly diseases

informatician (in-for-muh-TIH-shun) *noun* a person who analyzes medical information and helps determine the best care for patients

M

monkeypox (MUHNG-kee poks) *noun* a viral disease that causes fever, fatigue, and raised bumps on the skin

P

parasite (PA-ruh-site) *noun* an animal that has to live on another animal to survive

pathologist (puh-THOL-uh-jist) *noun* a scientist who studies the causes and effects of serious diseases

petri dishes (PEE-tree dish-uz) *noun* circular glass containers used to grow cultures of bacteria for investigations

R

rabies (ray-beez) *noun* a viral disease that causes drooling, problems moving, and eventually death

S

salmonella (sal-muh-NEL-uh) *noun* a disease caused by Salmonella bacteria. It is marked by food poisoning, stomach inflammation, and fever in humans and other mammals.

sterile (ster-uhl) *adjective* clear of any bacteria or viruses

strain (strayn) *noun* a specific version or type of bacteria or virus

symptoms (SIMP-tuhmz) *noun* the physical signs of a disease

T

toxoplasmosis (tok-soh-plaz-MOH-sis) *noun* a disease caused by *Toxoplasma gondii*. It is marked by muscle aches and pains, fatigue, and swollen glands.

transmitted (trans-MIT-ed) *verb* to have spread an infection from one person or animal to another

tularemia (too-luh-REE-mee-uh) *noun* a disease often spread by rabbits. It can cause fever, headache, cough, and diarrhea.

V

vaccine (vak-SEEN) *noun* an injection of a weakened or killed microorganism like a bacteria or virus. A vaccine is given to prevent or treat infectious diseases.

veterinarian (vet-ur-uh-NAIR-ee-uhn) *noun* a doctor who is trained to diagnose and treat sick or injured animals

virus (VYE-ruhss) *noun* a tiny germ that grows and reproduces in living cells; some can cause serious illness

Z

zoonoses (zoh-uh-NOH-seez) *noun* diseases that are passed from animals to people

zoonotic (zoh-uh-NOT-ik) *adjective* having to do with a disease that's spread from animals to people

Index

Author's Note

Y ou've finished this book, and you can't get enough of zoonotic diseases? Here's what I did to gather the information found in this book, along with a few tips for starting your own research on this topic.

When I was compiling this book's research, I started with the Internet. I used a search engine to look up terms such as "zoonosis" and "animal-borne disease." That led me to a wealth of information on this topic—everything ranging from complicated scientific journal articles to real-life stories of people affected by zoonotic diseases. I chose three of these true stories for the case studies in this book.

To get first-person accounts of these true stories, I had to leave the Internet behind. I called the three experts featured in the case studies—Dr. Melski, Dr. Mitchell, and Dr. Hammond—and made appointments to talk with them over the phone. Chatting with real people who work with zoonotic diseases gave me clear information and a personal perspective that I wouldn't be able to access any other way.

If you'd like to talk with real people about how they solve medical mysteries, including zoonotic ones, the best place to start is with the medical professionals you know. Family doctors and veterinarians have plenty of experience in diagnosing and treating zoonotic illnesses. Many of them would be happy to answer your questions and thrilled that you're curious enough to ask.

Before you chat with an expert, you can brush up on the basics the same way that Dr. Melski and other professionals do. Start your research online, in books at home, or at your local library.

Happy investigating!

ACKNOWLEDGMENTS

I would like to thank the following people for taking the time to help me gather the information necessary for this book. Without their assistance, this book would not have been possible: Dr. Roberta Hammond, Dr. John Melski, Dr. Kim Mitchell, Chris Schelpfeffer, Fernando Senra, and Maria Ascaño.

CONTENT ADVISER: Mark S. Dworkin, MD, MPH & TM, Associate Professor, Division of Epidemiology and Biostatistics, University of Illinois at Chicago